Abibe SENE

URBAN CHALLENGES AND AGROPASTORAL REQUIREMENTS

Abibe SENE

URBAN CHALLENGES AND AGROPASTORAL REQUIREMENTS

IN THE SAFÉNE REGION EAST OF THE SENEGALESE CAPITAL

ScienciaScripts

Imprint
Any brand names and product names mentioned in this book are subject to trademark, brand or patent protection and are trademarks or registered trademarks of their respective holders. The use of brand names, product names, common names, trade names, product descriptions etc. even without a particular marking in this work is in no way to be construed to mean that such names may be regarded as unrestricted in respect of trademark and brand protection legislation and could thus be used by anyone.

Cover image: www.ingimage.com

This book is a translation from the original published under ISBN 978-620-6-71532-0.

Publisher:
Sciencia Scripts
is a trademark of
Dodo Books Indian Ocean Ltd. and OmniScriptum S.R.L publishing group

120 High Road, East Finchley, London, N2 9ED, United Kingdom
Str. Armeneasca 28/1, office 1, Chisinau MD-2012, Republic of Moldova, Europe
Printed at: see last page
ISBN: 978-620-7-92079-2

Dr Abibe SÉNE

Specialist in Management and Development of Rural Areas (GDER)

URBAN CHALLENGES AND AGROPASTORAL REQUIREMENTS IN SAFÉNE COUNTRY EAST OF THE SENEGALESE CAPITAL

FOREWORD

During our early childhood, we'd get up early in the morning to accompany our father to the fields. On the damp grass, we'd sweep away the dew with our arms, heading for "Ngoltongo"[1] , barely able to see where, between the orchards and bushes, the cobwebs we'd pull out with our feet, or, at a certain height, tickle our faces. And during the dry season, we'd head out to the fields every evening with a bowl full of couscous to give the shepherd who was fertilizing the fields his dinner. As a farmer's son, our daily life was rhymed like this throughout the year, and everything revolved around farming and animal husbandry. During this period, these activities fed their men and were the dominant sector in the area. Nowadays, however, these lands cannot even be accessed because of the freeway, which has become an obstacle, and much of it is in the hands of investors and used for other purposes. The main cause of this situation is the sustained urbanization of the area, fronted by the capital, and manifested in the rapid occupation of the land. The State is the prime mover behind this phenomenon, with its attempt to relieve congestion in the capital, symbolized by the transfer of many companies and projects to the Dakar-Thiés-Mbour triangle. It all began in 2000, as soon as the Liberals came to power, with the installation of Blaise Diagne International Airport in Diass, a key element in the reconfiguration of the area. This situation is now in its twilight, as all the factors that encourage it are present in the area, and land reserves continue to be plundered to the detriment of traditional activities, forcing players in this sector to retrain or leave the area as a resilience measure. Given the area's potential as a receptacle for large-scale food self-sufficiency projects such as REVA and GOANA, should these activities be stifled in favor of unbridled urbanization?

---

1 The fields are located just beyond the toll freeway, closing in on highly fertile land and a large part of the land reserves of the Sindia commune, particularly the villages of Kiniabour1 and 2, Sorokhassap and Sindia. It is this land that provides the best yields in the locality, with significant millet and peanut production.

GENERAL INTRODUCTION

West African cities in general, and Senegalese cities in particular, are undergoing decisive change[2] . This change is affecting spatial and social dynamics, production processes, management frameworks and types. It was above all the advent of a liberal government in 2000 that favored "the explosion of the real estate market, with the arrival in force of private developers and speculators in a context of growing individualization and privatization of the channels of access to housing, increasingly refocused around the household"[3] .

Faced with such a situation, major projects have been multiplied: relocation of Léopold Sédar Senghor airport - a business district is planned on the current site - with the development of a new airport platform in the rural hinterland, at Ndiass, widening of the RN1 to 2 or even 3 lanes in places, development of the Dakar-Diamniadio-Thiès freeway[4] .

This policy of decongestion, symbolized by the relocation of businesses, infrastructure and certain criteria of urbanity, will push the limits of rurality to the east of the Cape Verde peninsula, more particularly in the satellite zone of the Dakar-Thiés-Mbour triangle. "This renewed interest on the part of the State in Senegalese cities is also in line with the process of decentralization and the transfer of urban management to the local level, encouraged by the World Bank in the 1990s"[5] . International institutions, notably the International Monetary Fund (IMF) and the World Bank (WB), intervene on a massive scale via resources in the policy agenda and often set the framework[6] .

---

[2] Piermay (J-L) et *al*, 2007, *La ville sénégalaise. Une invention aux frontières du monde*, Paris, Karthala, 242 p.

[3] Tall (A), 1999, "Croissance urbaine et stratégies résidentielles des ménages l'exemple des quartiers spontanés de Dakar", *L'Union pour l'Etude de la Population Africaine*, n° 39, p. 7.

[4] Diongue (M), 2010, *Périurbanisation différentielle : mutations et réorganisation de l'espace à l'Est de la région dakaroise (Diamniadio, Sangalkam et Yéne), Sénégal*, Thèse de doctorat (type de thèse) de Géographie, Université Paris Ouest Nanterre La Défence, p. 14-16.

[5] Diongue (M), 2010, op. cit. p.15.

[6] Sow (M), 2010, *Agglomération dakaroise au tournant du siècle, Vers une réinvention de la ville africaine*? PhD thesis in spatial planning - urbanism, Université Paris Ouest Nanterre La Défense, Ecole doctorale, économie, organisation et société, <u>academia.edu</u>, p. 25.

These cities, and the capital in particular, have a major influence on the restructuring of this area, considered to be a hub for the future of the Senegalese economy. Thus, "the Dakar peri-urban area is an ideal spatial framework for observing the seeds of tomorrow's city, a privileged observatory of urban manufacturing, mutations and innovations"[7] .

This will also lead to a drastic reduction in agropastoral areas and the decline of these traditional activities, with the arrival of investors looking for land to house their projects. As a result, extensive farming and livestock breeding are experiencing serious problems, with a considerable drop in yields, forcing its players to retrain to adapt to a new economy, supported more by the informal sector. This urbanization is accompanied by a diversification of the range of off-farm activities.

These findings raise the following questions: What is the impact of urbanization on extensive agro-pastoral activities? Is it not at the root of the vulnerability of these activities, causing a drastic reduction in sowings? Faced with this situation, what are the resilience strategies adopted by the local population?

The aim of this book is to analyze the impact of the area's infernal urbanization on agropastoral practices and the population's resilience strategies. It is divided into four chapters: the first, on methodology, describes the context and the tools and methods used for data acquisition. The second, entitled "The Dakar-Thies-Mbour triangle: an area in the throes of urban change", attempts to analyze the pace of urbanization in the area, focusing on the factors behind this phenomenon, as well as the land dynamics it has engendered. The third, entitled "Agropastoral constraints", shows the problems that urbanization of the area has caused, especially for extensive farming and livestock rearing. It focuses on the decline in acreage and production. And finally, the fourth section, entitled "Resilience strategies of the population", attempts to identify all the alternatives available to the population in terms of agropastoral and other activities, as well as their financial limitations.

---

[7] Diongue (M), 2010, op. cit. p. 14.

CHAPTER 1: BACKGROUND, THEORY AND METHOD

Apart from the context, which places this work in the context of globalization and the urbanization of West Africa in general and Senegal in particular, emphasis was placed on the infernal phenomenon that characterizes the capital, with 97.2% in 2012 according to ANSD[8] , but also the Dakar-Thiés-Mbour triangle and its many consequences.

It is also a question of setting up an itinerary and the necessary means for a rigorous and methodical analysis, likely to carry out this study well. As with any scientific work, it is important to carry out fieldwork with well-defined objectives, supported by hypotheses and a well-adapted methodology. For the sake of completeness, we have focused on the analysis of field sociologist Olivier de Sardan (1995)[9] , through the collection of qualitative data, with tools ranging from documentary reviews to focus groups, structured and semi-structured interviews. For quantitative data, the most convenient and common instrument is the questionnaire, which is characterized by its diversity, given the mass to be covered.

In the end, many households, resource persons, business leaders, associations, agropastoralists... within and outside the locality, were interviewed and many sub-themes reviewed through a plethora of questions, in order to take full charge of our theme.

1.1. Context

1.1.1. The global context

The current international context is marked by globalization, rapid urbanization and economic concentration, and the challenges of sustainable development in the context of climate change. Over the past two decades, we have witnessed the globalization of the world economy, accelerated by the development of Information and Communication Technologies (ICT). This globalization is accompanied, among other things, by a permanent break-up and relocation of production value chains. The countries that benefit most are those able to offer comparative advantages in terms of innovation, quality and

---

[8] National Agency for Statistics and Demography
[9] Olivier de Sardan (J-P), 1995, "La politique du terrain", Enquête [En ligne], 1 | 1995, online July 10, 2013, accessed October 15, 2022, France. URL : http://enquete.revues.org/263 PEDIDAS http://www.pdidas.org/fr.

cost of labor, volume and proximity of markets and infrastructure. In this context of globalization, each country positions itself according to its degree of competitiveness. To this end, a number of development and globalization integration tools have been introduced throughout the world. These mainly involve instruments aimed at creating and producing competitive goods and services, such as clusters, special economic zones, export processing zones and transport infrastructures such as ports and airports, which act as interfaces connecting different countries to world markets.

Globalization is also accompanied by economic concentration and metropolization. On a global scale, production is concentrated in large cities, dynamic provinces and wealthy countries. Half of the world's production comes from 1.5% of the planet's land[10] .

This situation is a corollary of galloping urbanization. Indeed, in 2011, for the first time in history, almost 50% of the world's population lived in cities. This trend is set to intensify, according to United Nations forecasts, which estimate that between 2000 and 2030, the surface area occupied by cities is set to triple[11] . Consequently, the territorial development approach could be a lever for mitigating urban concentration, but also a means of promoting development in different territories by drawing on their resources and potential.

Moreover, until recently, economic development was essentially based on the intensive exploitation of natural resources. With the increasing scarcity of natural resources, climate change and the emergence of civil society, the concept of sustainable development is gaining ground, with the aim of achieving economic development that integrates environmental and social aspects. It is a development model that "meets the needs of the present without compromising the ability of future generations to meet their own needs". Through various international treaties, Senegal is committed to integrating the principles of sustainable development into its various national policies.

1.1.2. Sub-regional and national context

---

¹⁰ World Development Report, World Bank, 2009
¹¹ World Urbanization Prospects, United Nations, April 2004

In the sub-region, Senegal is involved in structuring projects such as the Organization for the Development of the Gambia River (OMVG) and the New Partnership for Africa's Development (NEPAD). Additional Act No. 3 of January 10, 2004, adopting the UEMOA's community development policy, stipulates that "the community's regional development policy aims to build a stronger, more attractive and more competitive union, with a regional market in which each State optimizes its comparative advantages through complementarity".

This work comes at a time when Senegal, and particularly the capital, is undergoing a phase of sustained urbanization, with diverse and complex consequences that are constantly torpedoing it, ranging from promiscuity to insalubrity, insecurity, flooding, unemployment... Thus, the country's overflowing and unchecked urbanization, particularly of the capital, is constantly causing harm to these peripheral localities, which are considered to be the receptacle of new development policies and territorial balances. As a result, "the Dakar conurbation is home to 19% of the country's total population and almost 54% of the urban population, estimated at 39% of the total population"[12] .

This trend is confirmed by Houndechandji, who states that "for the country as a whole, the urban population is increasing annually by 4.5%, while for Dakar this rate reaches 7%"[13] . And according to ANSD/RGPHAE, the urban population of the Dakar region will rise from 3,026,316 in 2013 to 4,199,856 in 2025[14] . And this situation continues to be fed every day, given the "desert" that characterizes the rest of the country.

As a result, Senegalese authors such as Diong, 2010; Thiandoum, 2013; Tall, 1999; Pouye, 2003; Ndiaye, 2012, have emphasized the urbanization of this area, citing as key factors the influence of cities such as Dakar, with its urban frontage, and the provision of infrastructure to correct special disparities,

---

[12] Jacob (J-P) and Delville (P-L), February 1994, *Les associations paysannes en Afrique: organisation et dynamiques*, APAD-KARTHALA-IUED, p. 293.

[13] Houndechandji (M-F), 2012, *Urbanisation et gestion des ressources naturelles dans le domaine des Niayes à Dakar : impacts sociaux, environnementaux et sanitaires*, Thèse de doctorat de troisième cycle, UCAD, département de géographie, p. 10.

[14] ANSD (Agence Nationale de la Statistique et de la Démographie), RGPHAE, 2013, *Projection de la population des régions du Sénégal de 2013 à 2025,* www.ansd.sn.

piloted by the State on the one hand and since 2000 on the other. According to Ndiaye (2012): "One of the main players in this peri-urbanization is the State, which has made it a strategic site since 2000 with the implementation of major projects underway to relieve congestion in the capital"[15] .

On the other hand, the policies of decentralization and decongestion of the capital since 2000 have led to the installation of infrastructures to the east of the capital, pushing back the boundaries of rurality on a daily basis, with the installation of a highly captivating informal economy. This situation prompted the State to turn its attention to satellite areas such as the Dakar-Thiés-Mbour triangle. It was the advent of liberal government in 2000 that favored "the explosion of the real estate market, with the arrival in force of private developers and speculators in a context of increasing individualization and privatization of housing access channels, increasingly refocused around the household"[16] .

As a result, land dynamics and issues are becoming more complex, as infrastructures will be accompanied by a large foreign workforce and investors, increasing the demand for space for housing or investment, to the detriment of local people and their agro-pastoral practices.

1.2 Theory

The current situation in the Saféne region gives rise to a great deal of reflection, given its geographical position, which places it on the doorstep of the capital, with a very favorable territorial offer. While the urbanization of this locality is one of the State's priorities, the fact remains that food self-sufficiency, so much trumpeted by public policy, is virtually a mirage.

And it's worth remembering that this area has significant agropastoral potential, with highly fertile and available land that can help Senegal reduce its food deficits. The area has also been home to national projects such as

---

[15] Ndiaye (S), 2012, *État des Lieux de l'Environnementale et des Ressources Naturelles de la Communauté Rurale de Sindia,* Geography dissertation, UCAD, p.45.

[16] Tall (A), 1999, "Croissance urbaine et stratégies résidentielles des ménages : l'exemple des quartiers spontanés de Dakar", *L'Union pour l'Etude de la Population Africaine,* n° 39, p. 7.

REVA[17] and GOANA[18] , initiated by President Abdoulaye Wade as part of his policy of agri-food sovereignty. Apart from these, the area is home to many agribusiness projects, most notably large-scale market gardening, such as the Dutchman *Van Oers'* perimeters in Kiréne, and Charles ADDAD's arboriculture, more centrally located and dominated by *Agouillo-Sénégal*, in Kiniabour2 (in the commune of Sindia).

And nowadays, urbanization seems to be compromising this reality, depriving this locality of a great deal of land, which is highly suitable and guarantees many agropastoral prospects.

Should we sacrifice this potential for the sake of housing and infrastructure, putting this population in such a deleterious situation?

This study is based on the hypothesis that urbanization of the area has sharpened the land stakes, with a decline in agropastoral areas, prompting the population to adopt resilience strategies as a means of adapting to this new situation.

- Since 2000, the east of the Senegalese capital, and more specifically the Dakar-Thiés-Mbour triangle, has been undergoing a remarkable urbanization phase, a situation encouraged above all by an attempt to relieve congestion in the capital.

- This situation has sharpened land dynamics and issues, leading to a decline in traditional activities, with a significant reduction in areas suitable for sowing and grazing, as well as in agro-pastoral production.

- Faced with this situation, the mainly indigenous and agro-pastoral population is obliged to retrain as a means of resilience, given the boom in new economic prospects being identified in the area.

1.3 Methods and tools

Data was collected using a mixed method, combining qualitative and quantitative research. This work took us to four communes (Diass, Sindia, Popenguine-Ndayane and Keur Moussa), and apart from resource persons,

---

[17] Retour Vers l'Agriculture (Return to Agriculture), initiated by President Abdoulaye Wade in the 2000s, to promote agriculture and achieve food self-sufficiency by massively encouraging the population to return to farming. The locality of Kiréne in the commune of Diass was one of the pilot sites for this policy.

[18] Grande Offensive Agricole pour la Nourriture et l'Abondance (same agricultural policy as the **RVA**).

we interviewed a total of 307 households. The quantitative data mainly provided us with information on the number of households, and three questionnaires were used to corroborate this approach:

The first, entitled "The practice of agriculture", is addressed to farmers and comprises two parts: the demographic grid, similar to all the other questionnaires, comprises 13 questions and relates to the identification of the respondent, followed by the section on socio-economic activities. This questionnaire is structured around 57 questions and focuses on farming practices, i.e. agricultural production, land acquisition, constraints linked to this activity and the availability of areas suitable for sowing, without forgetting those sold. A total of 116 people (37.7%) were interviewed, with the most targeted localities being Thiafoura, Khassap, Bandia, Sindia Kaf Ngoune in the commune of Sindia, and Kiréne, Raffo and Samekédj in the commune of Diass.

The second, "The practice of livestock farming", also comprises two parts: the demographic grid and another reserved for socio-economic practices. This section contains 45 questions, and covers livestock farming in all its diversity, species, production, availability of water and space, and all the factors that hamper this activity, as well as solutions. Land tenure wasn't mentioned very much, as breeders in the area are generally foreigners and don't own land.

A total of 44 farmers (14.3%) were interviewed, and the most targeted localities were: *Khong khalma*[19] , Sindia Croisement in the commune of Sindia, *Altou Well*[20] , Samekédj, Bentegné, Tchiky and Togglou in the commune of Diass.

Finally, the third questionnaire is entitled "Non-agricultural practitioners" and also comprises two parts: the demographic grid and the socio-economic section. The latter covers all the extra-agropastoral activities taking place in the area, as well as their profitability and limitations. It shows the functional diversity of the locality. Among those interviewed, we note: industrial

---

[19] Hamlet located in the commune of Sindia, on the Thiés-Sindia departmental road, just outside Sindia on the Dakar-Baobab circuit. It is inhabited by Peulhs, whose main activity is livestock breeding.
[20] Contains the same features as *Khong Khalma*, except that it is located on the RN1 and in the commune of Diass.

employees, the handicraft sector, transport, mining and quarrying, shopkeepers, agricultural employees and all those whose income does not come from agropastoralism. The influence of urbanization in the area on their activities was also on the agenda. Emphasis was also placed on the sale of land, in order to estimate the importance of this phenomenon within the sector, and a total of 40 questions were used. In the *end*, 147 people were interviewed, representing 28% of our database: Sindia, Guéréo in the commune of Sindia, Ndayane and Popenguine in the commune of the same name, Diass, Kiréne, Togglou, Bentégné in the commune of Diass, the Sites de recasement in the commune of Keur Moussa.

The qualitative data are based on four modes of intervention, according to the principles of the socio-anthropologist Olivier de Sardan: documentary analysis, participant observation, formal or informal individual interviews and focus groups.

1.3.1. Document review

To be more contextual, a good part of the literature review was devoted to works that talk about locality, urbanization and peri-urbanization in general. Among those that contributed most to shaping this work were Diongue's thesis (2010)[21] , which discusses peri-urbanization in the east of the Senegalese capital. This work provides an exhaustive analysis of the facets of this phenomenon, with the interplay of actors, in a highly strategic locality. Thiandoum's dissertation (2013)[22] , which discusses agricultural inputs in a peri-urban locality (le pays saféne). She placed great emphasis on the potential of these activities, despite the changes the area has undergone. Also of note is ANAT (2015)[23] , through the master plan for planning and development of the Dakar-Thiés-Mbour triangle. This document seems to

---

[21] Diongue (M), 2010, *Périurbanisation différentielle : mutations et réorganisation de l'espace à l'Est de la région dakaroise (Diamniadio, Sangalkam et Yéne), Sénégal*, Thèse de doctorat (type de thèse) de Géographie, Université Paris Ouest Nanterre La Défence, p. 14-16.

[22] ANAT, 2015, Schéma directeur d'aménagement et de développement territorial de la zone Dakar-Thiés-Mbour, Ministère de la Gouvernance Locale, du Développement et de l'Aménagement du Territoire, Rapport provisoire, 163 pages.

[23] Thiandoum (M), 2013, *Espace périurbain et activité rurale: dynamiques territoriales en cours en pays saféne*, Thèse de Doctorat unique de Géographie, UCAD/ETHOS, 324 p.

exclude agropastoralism in the area, placing greater emphasis on urbanization and development.

All in all, these works have given us a great insight into the challenges of urbanization in the area, as well as its agro-pastoral potential, all symbolized by a major interplay of players.

1.3.2. Participatory observation

Olivier de Sardan defines participant observation as "the very heart of ethnographic fieldwork". Fieldwork is one of the biggest investments of time. Prolonged interaction with actors *in situ* (in their natural sites, in their natural living conditions) produces two types of effects:

- The first, the most visible, the most formal, is the field notebook, which records observations, eavesdropping, chatter, discussions, in a social flow of some kind.

- A second effect is equally important: impregnation, i.e. a whole series of informal processes by which an interviewer becomes accustomed to understanding all social codes and social logics of behavior at their most impalpable, everyday level. In order to grasp a certain number of social processes in their social and natural context.

Numerous visits were made both inside and outside the zone, to meet with stakeholders (agropastoralists, practitioners of other activities, administrative authorities, traditional chiefs, etc.), in all their ethnic, professional and cultural diversity...

1.3.3. Formal and informal individual interviews

Formal and informal interviews were our main information-gathering tools. In prolonged interaction, interviews tend to be as close to conversation as possible. It is a strategy considered central to the field anthropologist, to "leave the interview as close as possible to the natural forms of interlocution common in local society"[24] . Classically, we have distinguished two registers in our interviews: one in which we take the interlocutor as a "consultant", and one in which we take him as a "reciter". The consultant was asked to talk about the main causes of urbanization in the area, land and agro-pastoral

---

[24] Olivier de Sardan (J-P), 1995, Op. Cit, p.14.

changes and their consequences, as well as resilience strategies in the face of this situation, etc., as one would ask an expert (or a "resource person").

These resource persons included: the presidents of the state commissions of the target communes, local elected officials, company managers, village chiefs, livestock and agricultural organizations, development associations, etc. Some of them were consulted several times.

On the other hand, the interviewee was asked to testify using a biographical sequence, i.e. what he or she had experienced as a local resident, and his or her position with regard to the changes, hence the need for "life stories". The tool used is the interview guide, with specificity depending on the nature of the activity or responsibilities.

1.3.4. The focus group

This technique makes it possible to diversify stakeholder discourse in order to assess needs, expectations, satisfactions and/or better understand the meaning they give to their opinions, motivations and behaviors. "The focus group is a qualitative method of collecting data based on the discussions of a group of participants on a topic predefined by a facilitator.[25]

Three focus groups were carried out with workers (in factories, farms, quarries, the transport sector, commerce, etc.), and this work was made possible thanks to the managers of the sectors concerned.

The first focus, addressed to workers, attempts to discern the influences of changes in the area on their work, the constraints and advantages, the profitability of their activity, and its future. Factories, farms, quarries, garages... were our targets. The second was aimed at livestock farmers, and focused on pastoral constraints, production and alternatives in the event of a decline in their activity. The target areas are water points (boreholes, wells) and *loumas*[26] , where herders converge. Finally, the last session is aimed at farmers, and deals with the decline in acreage and production as a result of urbanization, new agricultural strategies and prospects, and the commercialization of land. The most targeted localities are those where

---

[25] Morgan (DL), 1996, Focus Group, Annual Review of Sociology, 22, 129-152, http://dx.doi.org/10.1146/annurev.soc.22.1.129.
[26] Local weekly market

agriculture continues to exist (Thiafoura, Samekédj, Kiréne...). The hours targeted were during meals under the palaver tree, when the farmers were off work, otherwise they were busy all year round.

Conclusion

At the end of this first chapter, the stage is set, and the itinerary for this work is clear. Like any scientific study, this one cannot do without a methodology and sworn tools for data collection. And to be exhaustive and comb through the hypotheses, it was necessary to use a mixed method, waltzing between quantitative and qualitative data. For quantitative information, the most appropriate tool is the questionnaire, which is characterized by its specific thematic diversity. Our qualitative approach was primarily guided by the work of field sociologist Olivier de Sardan, with an approach that we feel is comprehensive, making use of all the tools necessary to tackle this theme.

This method enabled us to take on various sub-themes, various populations, with various questions, justifying the importance of the subject, the scientificity of the work and the desire to provide plausible results for the credibility and scientificity of this work.

CHAPTER 2: THE DAKAR-THIÈS-MBOUR TRIANGLE: AN AREA UNDERGOING URBAN TRANSFORMATION

Urbanization is a phenomenon long observed in Africa, with multiple and complex causes. Cities are expanding, and the urban appears as a complex planetary reality, evoking the border between order and chaos, uncertain, unfinished and imperfect[27].

Sub-Saharan Africa thus occupies a special place in the urbanization of developing countries. This rapid urbanization reached its peak in the 1950s, when the annual growth rate of the urban population in agglomerations of over 10,000 hbts reached 7.1% per year. Between 1950 and 1990, the urban population of Sub-Saharan Africa increased tenfold, while the total population tripled[28]. In terms of infrastructure, this area has been planned by the State to host major structuring projects. "One of the main players in this peri-urbanization is the State, which has made it a strategic site since 2000, with major projects underway to relieve congestion in the capital"[29].

Thus, the opening of the AIBD in 2000 was like a license to occupy the Dakar-Thiés-Mbour triangle, stimulating the arrival of investors and the complexity of land issues. Many conflicts ensued, pitting several players against each other, with varying ambitions and above all motivated by profit.

2.1. Urbanization factors

Today, the Dakar-Thiés-Mbour triangle is a booming area, not only in terms of urbanization, but also economically, socially and environmentally, with major hubs of activity and attraction, once again demonstrating the State's determination to transform this locality into an international up.

Map 1: Spatial distribution of business parks

---

[27] Da Cunha (A) and Matthey (L), 2007, "La ville et l'urbain: des savoirs émergents", *Presses polytechniques et universitaires romande*, p. 25.

[28] Bocquier (P), 1999, "La transition urbaine est-elle achevée en Afrique?", *Chronique du CEPED*, n°34, pp. 1-4.

[29] Ndiaye (S), 2012, *État des Lieux de l'Environnementale et des Ressources Naturelles de la Communauté Rurale de Sindia*, Geography dissertation, UCAD, p.45.

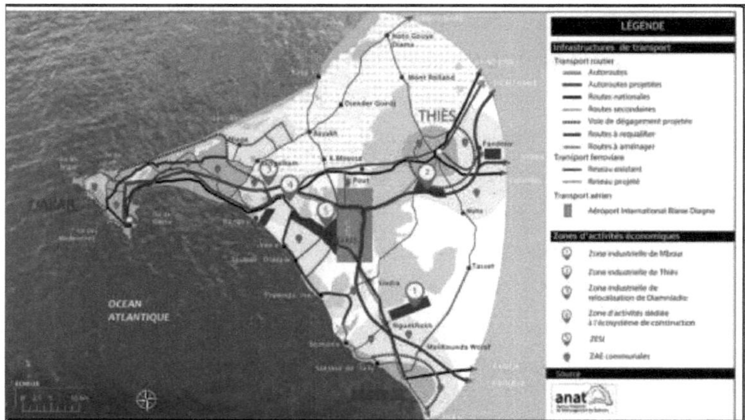

Source: ANAT, 2015

Projects such as the Blaise Diagne International Airport (AIBD), the Integrated Special Economic Zone (ZESI), the Ndayane mineral port, the urban centers of Diamniadio and Dagga-Kholpa, the Diamniadio industrial zone, Dakar's second university, and the Diamniadio-AIBD, AIBD-Thiès and AIBD-Mbour freeway projects, have encouraged the expansion of non-agricultural activities, devouring an enormous amount of land once used for agro-pastoral practices.

The triangle's considerable wealth and development potential, combined with Dakar's polarization, make it a strategic zone, as evidenced by the gradual installation of businesses and the construction of new housing. "The main obstacle to housing production is the absence of a structure in charge of land development other than the MUH and local authorities (since the dissolution of SCAT-URBAIN in 2000). The MUH has announced the creation of 25 urban hubs by 2035, including 5 in the Dakar-Thiès-Mbour priority zone"[30] . In view of all this, it is crucial to control and plan the development of this

---

[30] World Bank, 2015, IBRD-IDA, *Urbanization review: Emerging cities for an emerging Senegal,* 126 pages.

area. To avoid the syndrome of anarchic urbanization of Dakar, it is clear that the resources and potential of this triangle must be exploited to the full[31] .

In the industrial sector, many infrastructures have been built, the most important of which are: Ciments Du Sahel, with almost 350 ha and over 1,000 direct and indirect jobs, *Twyford Ceramics*, with 32 ha and its annex, employing over 1,500 people, and SIAGRO, which is active in agri-food production. It's worth noting that these infrastructures are attracting large numbers of people to an already densely populated area, as the 3 départements together represent 1.9% of the national territory and accounted for 13.5% of the national population in 2013. Indeed, the total population of the departments of Rufisque, Mbour and Thiès was estimated according to the provisional results of the latest general population census at 1,739,897 in 2013. With a total surface area of 3,852 km2, the average population density in this area is 453 hbts/km2 compared with an average density of 65 hbts/km2 at national level[32] .

Major state projects, whether announced or underway, create economic opportunity costs that have a real impact on the changing uses of peri-urban areas, making them more attractive[33] .

Figure 1 Factors in the urbanization of the area

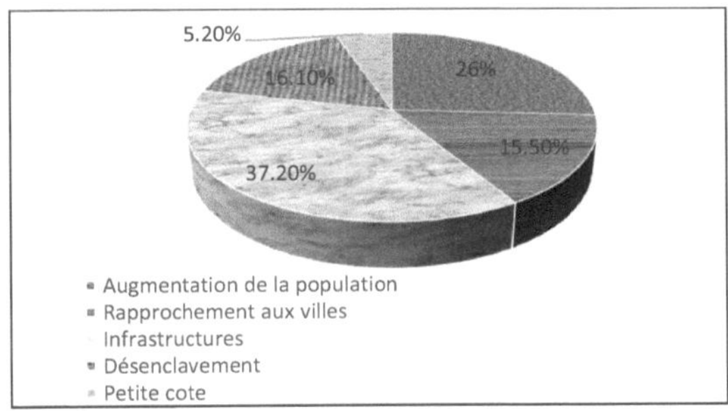

Source: Field surveys, 2022

---

[31] ANAT, 2015 (Agence National de l'Aménagement du Territoire), schéma directeur d'aménagement et de développement territorial de la zone triangulaire Dakar-Thies-Mbour, Rapport provisoire, p. 9.

[32] ANAT, 2015. Op. Cit, p. 31.

[33] Diongue (M), 2010, Op. Cit, p. 14-15.

As a result, 37.20% of the population believe that the presence of infrastructure is the main cause of the area's urbanization, versus 26% who believe that this phenomenon is due to its geographical position on the small coast.

"Since 2000, with the opening of the AIBD, other infrastructures have followed, and the locality receives foreign investment every day, providing employment for the local population. These projects have changed the structure of the area, and have cost us a lot of land for farming and livestock breeding"[34] .

This trend, far from having run its course, continues to sharpen the stakes in terms of land use, with an increase in the competitiveness of the locality, where players in all their diversity are driving the conquest of land, according to their ambitions and projects. Agricultural, industrial and residential areas make up a complex, often confusing mosaic, where land use is multiple and unclear. The management of these rapidly changing areas (...) is crucial and concentrates many of the development challenges facing the south"[35] .

2.2. The Diass hub: the government's desire to urbanize this triangle

The Diass hub straddles the communes of Keur Moussa, Diass, Sindia, Popenguine-Ndayane and Yéne. It currently comprises a network of semi-urban settlements (Popenguine, Sindia, Toubab-Dialaw, Diass, Yéne) located between the RN1 and the ocean, as well as a large number of villages and hamlets. It is structured around Blaise Diagne International Airport and the Integrated Special Economic Zone (ZESI). It is a larger-scale hub than Diamniadio, thanks in particular to the ZESI's potential for creating economic activities and jobs, and the availability of nearly 7,000 hectares of urbanizable land located on either side of the airport towards Pout and in the Dagga-Kholpa zone. The freeways currently under construction and the Dakar-AIBD rail line project will position this hub as a multimodal interchange hub for the conurbations of Dakar, Thiès and Mbour, while the AIBD and ZESI are

---

[34] Semi-structured interview with S. SENE, former president of the domanial commission of the Diass commune (site of the current airport), conducted on September 10, 2021.

[35] Dauvergne (S), 2010, *Dynamique des agricultures périurbaines en Afrique sub-saharienne et statuts fonciers, le cas des villes d'Accra et Yaoundé* : une approche de l'intermédiarité en géographie, PhD Thesis in Geography, Defended 08-12-2011 in Lyon, École normale supérieure, p. 61.

globalization interfaces that will open up this territory and Senegal to the world.

Based on current average population density in the departments of Dakar, Guédiawaye and Pikine, the Diass hub could eventually accommodate over 900,000 inhabitants.

At present, economic activities in this area revolve mainly around quarrying and tourism, which is developing along the small coast between Yéne and Toubab Dialaw.

If this hub is to fulfil its urban and economic function and be a territory of influence, contributing to a better structuring of the area, it must undergo fairly rapid but sustainable social and economic change, in order to welcome new populations and economic activities.

Map 2: Diass hub

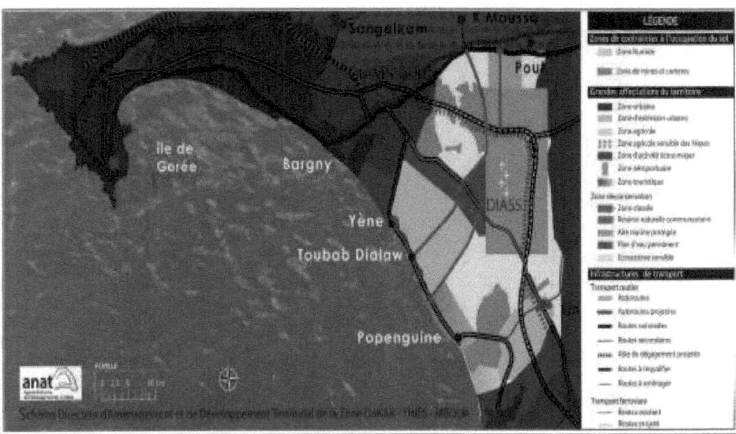

Source: ANAT, 2015

2.3 Multiplying land issues

All these infrastructures, combined with private investment, have led to changes in many areas, with galloping urbanization of the area and, above all, the development of housing to the detriment of traditional and agro-pastoral practices.

"The economic crisis of the 1980s put an end to the state's ambition to promote modern housing for the majority of city dwellers. Faced with financial constraints, the State resigned from its role as builder. Land production was therefore liberalized, with the entry of real estate companies and private individuals"[36] .

According to Pouye (2003)[37] , nearly 20 sites are being subdivided, according to S. SENE, chairman of the Diass rural community's land commission. SENE. These sites are located mainly along Route Nationale 1 (RN1), from Kandam to Bandia, where over 16,000 plots have been allocated for housing. This situation will sharpen the stakes in land tenure, with its high level of speculation leading the population to sell off their land, for fear of losing it to major government projects. This scramble for land, caused above all by urbanization, involves many competing players (natives, foreigners, entrepreneurs, private individuals, the state, etc.), under the arbitration of the authorities, who have turned into veritable rentiers. "Today, it's no longer the peri-urban era of the first encounters, when rural dwellers could claim a certain land legitimacy linked to their autochthony, and city dwellers a feeling of domination over the former"[38] .

Table 1 The land market in the area

Vente des terres	Fréquence
Population ayant vendu	32,6%
Population n'ayant pas vendu	67,3%
Total	**99,9%**

Source: Field surveys, 2022

Land is no longer given away, but sold. This new type of suburban land fragmentation represents a social break with the previous pattern of irregular insertion"[39] . Faced with this situation, some 32.6% of the population have sold off their land, either for fear of losing it to major government projects, or to meet urgent family needs. The localities most affected are Thiafoura,

---

[36] Djah (A-J) *et al,* 2017, "Les implications socio-économiques et spatiales des pratiques foncières des autorités coutumières dans le développement de la ville de Yamoussoukro (Côte d'Ivoire)", *Regardsud,* N°2, p. 163.
[37] Pouye (I), 2003, *La communauté rurale de Diass : Etude géographique,* DEA de Géographie, encadreur, Faculté des Lettres Sciences Humaines, UCAD/Dakar, p.12.
[38] Diongue (M), 2010, op. cit. p. 16.
[39] Bertrand (M), 1994, *La question foncière dans les villes du Mali, Marchés et Patrimoines,* Publié avec le concours du CNRS, KARTHAL-ORSTOM, p. 11.

Khassap, Kiniabour, in the commune of Sindia, Diass, Thicky, Kiréne, in the commune of Diass and to a lesser extent in the commune of Popenguine-Ndayane, given their low land reserves. This trend is weakly observed in the commune of Keur-Moussa, because land is not as valuable as in the preceding communes, which have received a large proportion of public and private investment.

The sale of land is not provided for in land legislation, but it has become common practice in Senegal. Indeed, "under pressure from foreign financial bodies such as the World Bank, which advocated legislative reform to develop the private sector, Senegal's 1996 land action plan focused almost exclusively on land privatization [...], while the practical needs of small family farming units were barely taken into account"[40].

In a certain logic, it is necessary to enable agropastoralists to own land in the National Domain (DN) to enable them to practice their activities freely, but also to settle there permanently and be able to assert their title deeds to banks, for example, to obtain credit. What's more, "land issues are becoming progressively more complex due to the increasing monetization of land value brought about by urbanization [...]"[41]. In Senegal, for example, DN land, once registered[42], enters directly into the private domain of the state[43].

This recent situation of land sales has been accentuated by the strategic position of the area, where "a market logic is developing with the promotion of speculative mechanisms"[44]. And in the peri-urban area of Diass and Sindia, land speculation is booming. "Land is being haggled over, and indigenous populations, mainly farmers and stockbreeders, are beginning to give up their fields for the money of private individuals and property

---

[40] Mayke (K), Yaram (G) and Marieke (K), (2013), "Les conflits fonciers au Sénégal revisités: continuités et dynamiques émergentes", in Gerti Hesseling, A l'ombre du droit, p.38.

[41] Mayke (K), Yaram (G) and Marieke (K), (2013), Op. Cit. p.39

[42] A procedure provided for by decree n°=64-573 in accordance with articles 8 and 11 of the law on the national domain in Senegal.

[43] Ka (I), 2018, Enjeux et défis de la gouvernance foncière en Afrique: contribution à l'étude du droit foncier à partir d'une analyse de sécurisation foncière rurale au Burkina Faso et au Sénégal, PhD Thesis in Law, UGB de Saint Louis, p.137.

[44] Diop (M), 2016, *La contribution des programmes immobiliers du pôle urbain de Diamniadio dans la résolution de la crise du logement à Dakar*, ESEA ex ENEA, Mémoire de master 2, ESEA ex ENEA, Aménagement du territoire, environnement et gestion urbaine (ATEGU), p.122.

developers"[45] . This dynamic of land liberalization can encourage land grabbing, marginalizing the most vulnerable sections of society[46] , as well as their traditional activities, for lack of space. "From a moral and legal point of view, land grabbing can also consist of arbitrarily dispossessing an individual or a community of their rights of enjoyment or cleanliness over the land they occupy"[47] . This phenomenon is more prevalent in the communes of Diass and Sindia.

As a result, we're witnessing a drastic reduction in agropastoral areas, leaving the players in this sector in a perplexing situation, with a vertiginous drop in production, as well as numerous and complex conflicts, involving many players.

2.4. Conflicts arising from land rents

The changes in land ownership in the area in recent years have had an impact on relations between the local population, as well as between investors and the authorities.

Figure 2: Types of land conflicts

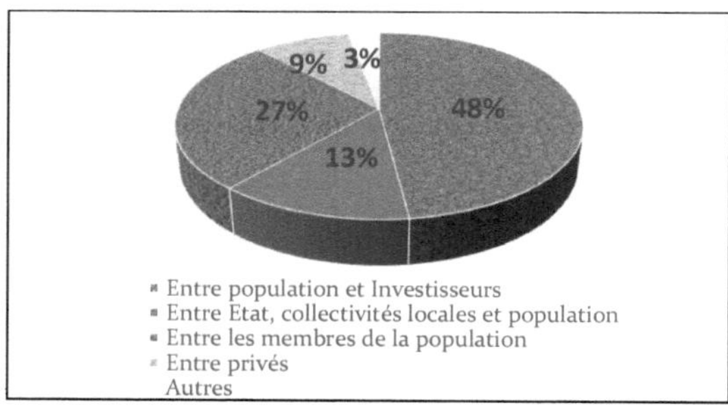

Source: Field surveys, 2022

These changes, which have led to an increase in land rents, have raised the stakes in this area and led to serious conflicts between players in the sector. According to Thiom, 18.50% of the population in this locality were involved

---

[45] Seck (S), 2011, *Analyse des mutations foncières au Sénégal: CAS de la communauté rurale de Diass,* Master 2 thesis, UCAD, geography department, GDER, p. 59.
[46] Puépi (B), 2015, Les gouvernances foncières et leur impact sur le processus de développement : Cas de quelques pays africain, Paris, L'Harmattan, p.189.
[46]

in land disputes in 2004, compared with 58% in the previous year, and this trend is set to increase and even reverse in 2015, with 68.80% of the population involved in land disputes compared with 5.60%[48] . These land conflicts are of various kinds and involve several players with different and complex ambitions, most of which are motivated by money.

2.4.1. Between the public and investors

Analysis of this graph shows that a large proportion of conflicts involve investors and the local population, accounting for 48% of cases. The most remarkable situation is that of Lebanese businessman Charles HADDAD against the population of Kiniabour2 , in the commune of Sindia, over an area of 200 hectares, destined for agro-business. It should be noted that this locality has a very good territorial offer, with land highly suitable for agricultural practices. The investor's installation on the site that L. Pouye[49] calls "the lung of Sindia", is in conflict with the local population, and the case is still before the courts.

2.4.2. Between local authorities and the general public

Between the State, local authorities and the population (13%), taking into account the major State projects raging in the area. The most affected localities are the communes of Keur Moussa and Diass, with the displacement of two villages (Mbadatt and Kathialique). The airport found a site already developed for housing, and the commune was not even informed of decree no. 2010-894 of June 30, 2010, which doubled the surface area of the AIBD from 4,000 to 8,000 hectares. "The state didn't warn anyone, even us local authorities. The local people are not going to stand idly by, because this area has already been allocated to certain people, so it's all the more reason to agree to compensate them with dignity"[50] .

2.4.3. Within the population

---

[48] Thiom (S), 2017, Analyses des mutations foncières dans la commune de Sindia, Master 2 thesis, Geography, FLSH, UCAD, 118 pages.

[49] Resident of Kiniabour 2 and member of the collectif pour la Défence des terres de cette localité, interviewed on November 22, 2022.

[50] Semi-structured interview with Assane SENE, president of the Diass commune's domanial commission, on November 22, 2022.

These conflicts can also be observed within the population, particularly within families (27%), as a lot of money is at stake, leading to the breakdown of parental ties. It should be noted that, despite the development of the land registration system, land management remains traditional, with land being passed down from father to son, and management being entrusted to the eldest member of the family as soon as the father dies. The latter is subject to two sharing options: by charikha (two for the man and one for the woman) or in court (equitable sharing between the two). This is not often the case, given the economic stakes involved in land ownership, and this phenomenon is at the root of conflicts and the break-up of many families.

2.4.4 Private parties

Better known as double allotment, this phenomenon is becoming increasingly recurrent in the locality, accounting for 9% of conflicts. In recent years, its frequency can be explained by the attraction of the area, symbolized by the superiority of demand over supply, in terms of land distribution. It is more orchestrated by the State, through local authorities, where one plot may be subject to several allocations. This phenomenon is more noticeable in highly urbanized areas such as Diass, Sindia, Bentégné and Guéréo, where land reserves are beginning to run out. It should be noted that many of these deals are motivated by money, and many cases in the area are pending before the courts.

Conclusion

The sustained urbanization of Pays Saféne has multiple and complex causes, including its geographical position, which places it at the lips of the capital and between two cities: one administrative (Thiés), with its demographic weight constantly expanding its limits, and the other tourist and seaside resort (Mbour), which continues to receive foreigners and investors on a daily basis, given its favorable territorial offer. The other remarkable phenomenon, which is reshaping the locality, is the development of infrastructure to relieve congestion in the capital. These factors have pushed back the boundaries of rurality, sharpening the stakes on land and making the locality a target for investors. This profitability of land is not in its twilight, as all the factors favourable to its development are becoming stronger every day in the area.

As a result, many players are involved, with the selling off and cornering of land resources orchestrated respectively by the local population and the State through its branches. This has led to a drastic reduction in agropastoral areas, putting these activities in a perplexing situation, very close to decadence in favor of a growing informal economy.

CHAPTER 3: AGROPASTORAL CONSTRAINTS

The advance of the urban front will lead to an automatic decline in the area under cultivation in favor of housing and urbanization, with very serious consequences for extensive farming and livestock breeding. As a result, "agriculture is also moving slowly [...]. The growth of the urban area is leading to a decline in cultivation [...]. Close to the city, the fields are retreating in the face of the urban thrust, blending in with the buildings"[51] . This is clearly perceptible to the east of the Senegalese capital, where a large part of the land and traditional activities are undergoing changes in the direction of modernity.

In context, this situation began in the 2000s, with the arrival of the Liberals in power, leading to the liberalization of land ownership and, more specifically, to the construction of the Diass Blaise Diagne International Airport. This infrastructure is both the cause and the effect of the conquest of the area's land, since its installation is budget-intensive in terms of land, and subsequently continues to encourage the installation of other infrastructures and populations. This situation, which continues to go from strength to strength, continues to reduce land reserves, much to the delight of informal activities. This in turn has led to a drop in agro-pastoral production, leaving the population in a perplexing situation of dependence, especially on food.

3.1. Decline in areas suitable for sowing and grazing

"Following the advance of the urbanization front, making land speculation an economically more powerful issue than agricultural speculation, areas formerly used for market gardening and fruit growing are being developed and transformed into residential estates and low-cost housing"[52] .

As a result, Doriel believes that peripheral agricultural land is being nibbled away daily by urbanization and the great retreat of traditional activities[53] .

This phenomenon, observed in many African cities, is not new, and continues to progress, to the great advantage of housing and infrastructure development,

---

[51] Chaleard (J-L) and Dubresson (A), 1999, *Villes et campagnes dans les pays du Sud : géographies des relations*, Paris, Karthala, p.59.

[52] Diop (A), 2011, *Aménagement en Zone littorale et risques industriels : l'espace Thiaroye sur mer-Mbao,* PhD thesis, UCAD, FLSH, Department of Geography, 332 p.

[53] Doriel (A-É), 2006, *Les Ires grandes villes dans l'urbanisation*, Institut français de l'urbanisme, p.76.

and to the detriment of rural life and extensive agropastoral practices. "Between 1960 and 1974, these areas shrank by a third, from over 27 million m^2 to nearly 9 million m ^{2} [54].

This urban push will increase demand for land in the industrial sector (757 ha), in intensive and peri-urban agriculture (1871.5 ha), much more noticeable in the center, in mining (92 ha), forestry (14337.35 ha), infrastructure and other sectors (23,157 ha), not to mention housing sites and traditional farms, the figures for which are informal. Together, these sectors have cost the area almost 40,214.85 ha of land, or around 70% of the capital's surface area. This clearly confirms the alienation of land in the eastern part of the capital, given the requirements and criteria of urbanity it hosts

Table 2 Agropastoral areas in the zone

Tendances observées	Fréquence
Baisse des surfaces	59%
Pas de baisse	41%
Total	**100%**

Source: Field surveys, 2022

As a result, "the land around and within the Niayes, much of which used to be devoted to agriculture, is rapidly being converted from permeable surfaces to meet housing and other urban infrastructure needs"[55] . This decline in surface area is observed by 59% of the population, compared with 41%, and the main cause is the sale of land and major government projects, which continue to encumber the area.

In the livestock sector, the main problem is the lack of species, reported by 35.2% of the population. The extensive practices that predominate in the area require a great deal of space to allow animals to roam and graze, while at the same time contributing to the diversification of biodiversity through the convoying and propagation of propagules via their fur or hooves. This lack of space is the cause of conflicts (23.5%), most often noted between farmers and herders, who share and compete for space every day, with complementary

---

[54] Gueye (N-F-D) et *al*, 2009, *Agriculteurs dans les villes Ouest africaines ; Enjeux fonciers et accès à l'eau,* IAGU-KARTHALA-CREPOS, p.191.
[55] UNDP, 2014, Strengthening the resilience of urban agricultural systems: assessing urban and peri-urban agriculture in Dakar, Senegal, https://attachment.outlook.live.net, 59 pages.

activities on the one hand, and antagonistic ones on the other. "Today, pastoralists are often seen as dangerous neighbors whose herds pose a permanent threat to crops. Cohabitation often turns into deadly confrontations: pastoralists then become adversaries rather than partners"[56] .

Table 3: Breeding constraints

Contraintes	Fréquence
Manque d'eau	17,6%
Manque d'aliment	18,4%
Manque d'espace	35,2%
Conflits avec autres acteurs	23,5%
Autres	5%
TOTAL	**100%**

Source: 2022 field surveys

This favors the bocage system in the area, a means for farmers to secure their properties and crops, using traditional hedges, supported by *balzami féra*[57] . Thus, according to Faye, "mobile livestock farming appears threatened in Senegal by climatic crises, agricultural pressure, the intention of protected areas, urbanization and the imported consumption patterns that accompany it, or by public policies favorable to its intensification and sedentarization"[58] .

This phenomenon is very common in localities such as Bandia, Kiniabour, Samekédj, Khassap, with a strong presence of mango cultivation, and in Samekédj, Raffo, Tchicky for manioc cultivation, which require protection and over a long period of time, to contain roaming herds, especially during the dry season. According to A. Alpha from Sindia: "The passage intended for herds (oxen, goats, etc.) at the entrance to Sindia coming from Khong Khalma to drink at the borehole used to be around 30 m long. Today, however, it has been broken up into plots and is only 10 m long"[59] .

For around half a century, Fulani pastoralists living in Sudanian agricultural zones have had to contend with the demographic growth of farming

---

[56] Marty (A), 1993, "La gestion des terroirs et les éleveurs: un outil d'exclusion ou de négociation", *in* Revue Tiers-Monde, Paris, PUF, T.XXXIV, n° 134, 327-344.

[57] A local plant better known as "Salaan", with an abundant, whitish sap, used to demarcate family properties and protect crops from straying animals.

[58] Faye, (B), 2006, "Les pastteurs sont des éleveurs " contemplatifs ". In Courage G, l'Afrique des idées reçues. Paris: Belin, coll. "Mappemonde", 399 pages.

[59] Semi-structured interview with the president of the Sindia commune breeders.

populations, the development of cash and food crops, and the reduction or disappearance of fallow land: deprived of their grazing lands in agricultural zones[60] .

Photo 1: Delimitation of Yérim Sow's property[61]
In Thiafoura (Nomboo)[62]

Photo 2: Herd stroking the fence
En grillage by Yérim Sow

Source: Field surveys, 2022

With individuals, and to a lesser extent, because the latter, motivated by profit, secure their perimeters to maximize their production. This leads to the

---

[60] Bernus (E) and Boutrais (J), 1994, Crises et enjeux du pastoralisme africain, Géographes de l'ORSTOM, département MAA, 213 rue La Fayette, 75480 Paris Cedex 10. CR. Acad. Agrlc. Fr, 80, n° 8, p.109.
[61] Yérim Sow is an Ivory-Senegalese entrepreneur born on April 23, 1967 in Dakar. He founded and presides over the Teyliom Group, active in 16 countries in Africa, Europe and the Middle East through 52 companies.
[62] Very fertile land, home to peanut, millet and corn crops, in the Thiafoura locality. They separate this locality from the communes of Nguékokh and Somone.

compartmentalization of grazing areas, resulting in a lack of feed (18% of the population) and natural water sources (17.6% of problems) such as "Somone", in the commune of Sindia and "Noungouma", in the commune of Diass, which apart from boreholes, are sources of fresh and brackish water for livestock.

This situation has led to a sharp drop in agricultural production from 2000 to the present day, and the partitioning of grazing areas and natural watering points as a result of certain developments, leading to indignation among the agro-pastoral population.

"Idrissa SECK[63] wants to dry up the *kaal*[64] , but we won't let him, because a good proportion of the area's herds drink here, and in the clear of the one the population goes there to catch brackish fish. So he has to free up the right-of-way of this watercourse for his rainwater supply"[65] .

3.2. Decline in surface area

To get a clearer picture, we've tried to periodize this evolution, taking as our starting point the installation of Blaise Diagne International Airport in 2000. A megaproject of this scale has reshaped the area, giving greater value to the land.

Figure 3: Evolution of land rent before 2000 to the present day

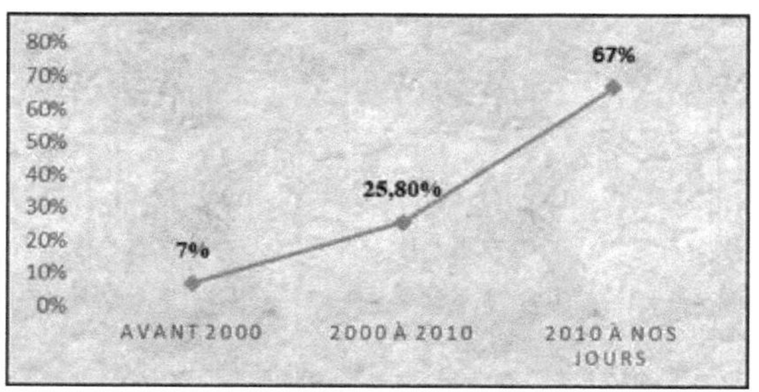

Source: Field surveys, 2022

---

[63] Politician, former prime minister under President Abdoulaye Wade and current president of the Rewmi party. He owns a private property in the area, whose installations interfere with the flow of the Somone river.

[64] Local name given to the "Somone" river, which rises in Diobass and flows through Nguékokh, Sindia, Khassap and Thiafoura, before emptying into the sea at Somone.

[65] Semi-structured interview with Daouda Dione, notable of the village of Sorokhassap in the commune of Sindia, conducted on September 10, 2022·

3.2.1. The situation before 2000

This period coincided with the experimental phase of decentralization policies and the transfer of skills to local authorities. New skills included land management by communes and rural communities. A new perception of land management was emerging, combined with the law on the National Estate, giving more power to the State to the detriment of the local population. The locality was not as attractive as it is today. Only 7% of the population had sold land, because people weren't interested in the area, and the infrastructures it housed were little more than their mere expression. As a result, much of the land was destined for farming and livestock breeding, and was neither alienated nor multi-purpose in its use. As a result, the land was available, and the land held by each farmer was more than sufficient for farming.

3.2.2. The situation from 2000 to 2010

This period was marked by the construction of the AIBD, one of the key elements in the reconfiguration of the area. This infrastructure contributed enormously to the occupation of space in the Saféne region, and was to be accompanied by other related structures, draining masses. During this period, the locality became increasingly attractive to investors, who saw it as a niche for their activities and a strategic point for the State's development policies. This sharpened the land stakes, with 25.80% of the population having sold land. It was during this period that the local population began to realize the future of the area and the land maneuvers to which it could fall victim.

3.2.3. The situation from 2000 to the present day

The stage was set by the AIBD, and will be set by countless public and private investments in the area. These infrastructures, whose installation was motivated by the locality's territorial offer, are in turn a real source of attraction for masses of people, most often workers and investors. This period was the turning point in the land mutations observed in the Saféne region, with the constant waltzing of investors and the State, with actions punctuated by heavy investments such as the Ndayane mineral port and the Dagga-Kholpa urban hub, all of which cost a great deal of money. Frightened on the one hand by state projects that swallow up huge areas of land, with lump-sum compensation payments that are borderline insignificant, and tempted on the other by investors, with a lot of money at stake, the local population had sold off a large part of their land. During this period, almost 67% had sold. This situation led to a considerable decline in the area under cultivation, followed by a drop in yields, putting these activities in a very complex situation and very close to decline.

3.3. A drop in production

3.3.1. Agriculture

The decline in agropastoral areas will reciprocally lead to a drop in production, as extensive practices are not highly mechanized and require a lot of space and manpower to ensure good production. This precarious situation will prompt some herders (38% of the total) to leave the area to avoid conflicts with private individuals and farmers, caused by the lack of space, by migrating to Joal or Djolof, where their herds can flourish.

Table 4: Agricultural production before 2000 to the present day

Période / Prod en t	Avant 2000	2000-2010	2010 à nos Jrs
[0-2]	12%	73,2%	88,7%
[2-4]	77,5%	23,2%	10,3%
[4-6]	8,6%	2,5%	0,86%
[6 et Plus]	1,7%	0,8%	0,0%

Source: Field surveys, 2022

An analysis of this table reveals significant variations in agricultural production from before 2000 to the present day:

- The highest yields of 6 tonnes or more were observed before 2000, with 1.7% of the population, before declining from 2000 to 2010, with 0.8% of users, and then levelling off from 2010 to the present day.
- The lowest yields of [0-2] tonnes were infinitely represented before 2000, with 12% of farmers. This population will increase, confirming the decline in production between 2000 and 2010, with 73.2% of the latter, compared with 88.7% from 2010 to the present day.

The bottom line is that these two extremes show that when yields are low, the population concerned becomes increasingly large, while the opposite is true when yields are high (6 tonnes or more).

This situation, which is worsening on a daily basis, puts those involved in agropastoralism in a very unfavorable position, forcing them to change their approach as a means of resilience, given the uncertain future of these activities. Subsistence farming is tending to disappear in favor of gardens and buildings[66] .

3.3.2. At pastoral level

Like agriculture, livestock farming will be affected by the reduction in acreage, forcing many farmers to leave the area or give up their activity. It should be noted that pastoralism in the area is still dominated by three species, with the cattle sector forming the backbone, accounting for 41% of the workforce. It is often coupled with goats, which account for 39%. The association of these two species is perceived more or less universally by practitioners, and is justified by the ability of the animals to adapt to harsh climatic conditions (heat, drought, lack of water, etc.), edaphic conditions (poor soils, lack of forage) and the long distances they have to cover. The sheep sector remains weak at 8.5%, given the difficulties mentioned above.

Figure 4: Dominant livestock species

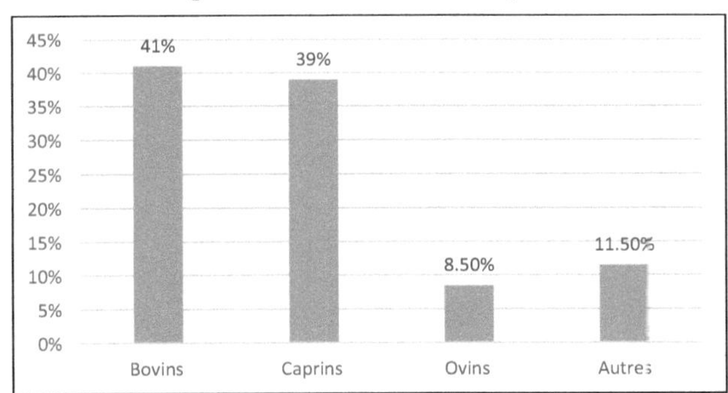

Source: Field surveys, 2022

Despite the dominance of the cattle sector, herd numbers are and continue to be limited, with low stocking rates per hectare, and herds rarely exceed one hundred head. The dominant herds are between [30-60] head, with 57.8% of the population, compared with 2.6% for those over 90 and around 100 head.

---

[66] Ba (A) and Moustier (P), 2010, Perception de l'agriculture de proximité par les résidents de Dakar, Revue d'Économie Régionale & Urbaine, N°5, Edition Armand Colin, pages 913-936

This phenomenon can be explained first and foremost by the difficult natural conditions, but also by the ever-tightening stranglehold placed on this activity by anthropogenic demands. While drought has led to a reduction in vegetation in many Sahelian countries, human development practices are causing a real imbalance in ecosystems. The development of buildings will lead not only to the partitioning of rangelands, but also to their reduction. This can lead to the loss of animals through malnutrition or disease.

Table 5: Cattle sector workforce

Nombre de têtes	Fréquence
[0-30]	13,1 %
[30-60]	57,8 %
[60-90]	26, 3 %
[90 et Plus]	2,6 %
TOTAL	**99,8 %**

Source: Field surveys, 2022

This low level of livestock numbers will lead to a reduction in meat and milk production. As far as meat consumption is concerned, we don't have the exact values to quantify this production, as a large proportion of the stalls are fed by animals that don't come from the area, and pastoralism in the locality, according to Faye, is "contemplative"[67] . Dairy production occupies more of the population, especially women, and varies according to the season. The figures obtained once again demonstrate the limits of this activity, despite its financial contribution to households.

Table 6: Milk production during the year

Quantité de production de lait en litres			
Pendant la saison des pluies		Pendant la saison sèche	
[0-5]	18,9%	[0-5]	81%
[5-10]	54%	[5-10]	18,9%
[10-15]	27%	[10-15]	0,0%

Source: Field surveys, 2022

The rainy season is the period of grace, given the availability of grass and water, and the dominant production is between [5-10] liters, with 54% of the

---

[67] Faye (B), 2006, "Les pastteurs sont des éleveurs " contemplatifs ". In Courage G., l'Afrique des idées reçues. Paris: Belin, coll. "Mappemonde", 399 pages.

population. Production between [10-15], considered to be the best, represents only 27% of this mass, and much of this production is sold in Dakar. During the dry season, there is a great regression, with a large proportion of production between [0-5] liters, i.e. 81% of the population, and quantities between [10-15] liters are non-existent.

It should be noted that this type of farming relies solely on grass cover and vertical fodder, and the animals run great risks during this period due to the lack of feed supplements. This often leads to the decimation of herds and a reduction in numbers.

Conclusion

For several decades now, the crises facing African agropastoralism have not been simply climatic or technical in nature, but have become social and political, to such an extent that the future of these activities seems doomed[68] . This is clearly perceptible in the Dakar-Thiés-Mbour triangle, particularly in the Saféne region, where since 2000, agropastoralism has been living to the rhythm of exacerbated urbanization, pushing back the boundaries of rurality on a daily basis.

This situation, initially triggered by the State through its policies to decongest the capital and regularize territorial inequalities, will be materialized by the installation of a great deal of infrastructure. This will subsequently be supported by the private sector, which is peddling investments all over the locality. This situation will have a terrible impact on land ownership, as much of it (fertile land) will leave the hands of agropastoralists to the benefit of investors and foreigners.

As a result, land reserves fell drastically, putting these activities in a complicated situation. This in turn led to a reduction in production from 2000 to the present day, forcing many of these players to leave the area or their activities as a resilience measure.

---

[68] Landais (E) and Lhoste (PH), 1990, "L'association agriculture élevage en Afrique intertropicale", *in* Sociétés pastorales et développement" Paris, Cah. Sc. Hum, ORSTOM, 26, 1-2, 217-235.

CHAPTER 4: POPULATION RESILIENCE STRATEGIES

Extensive farming and livestock breeding have long been the staple activities of the population, accounting for almost 52% of the total. These practices satisfied food and financial needs, especially in the event of excess production. In other words, a large part of the population limited themselves to satisfying their primary family needs, but in the event of surplus production, peasants marketed it to relieve some of their financial worries.

Today, these activities have lost their authenticity in terms of surface area employed, workforce, quantity of production and usefulness. This phenomenon is particularly encouraged by the changes that have taken place in the area, especially those relating to land tenure, which have given more space and power to investors to the detriment of agropastoralists and natives. Disadvantaged and deprived of their land, agropastoralists are seeing their activities slip away from them, and the few pockets of resistance are finding alternatives to maintain their practices. At the same time, almost 48% of this population are thinking of leaving this sector to invest in other areas as a way of adapting, given the rise of new trades in the area. And what will be the place of this population, which is neither well prepared nor trained to integrate the system that is taking shape?

4.1. Agropastoral

4.1.1. In the agricultural sector

In terms of agriculture, farmers tend to grow cash crops to supply the surrounding towns. "The influence of the city on the periphery is characterized by the flow of people in search of food products to satisfy the urban market. The periphery constitutes a green belt for the city's food security"[69] .

In view of this situation, the players in this sector are gradually moving away from food and cereal crops, "In a locality conducive to agricultural practices [...] on the edge of the massif"[70] . The same observation applies to most of Africa's peri-urban areas, where extensive livestock rearing and rain-fed

---

[69] Ngana (F) et *al*, 2010, *Transformations foncières dans les espaces périurbains en Afrique centrale soudanienne*, p.10.

[70] Demoulin (D), 1970, *Étude géomorphologique du massif de Diass et de ses bordures (Sénégal occidental*, PhD thesis, 3rd cycle in Geography, UCAD, p.37.

agriculture are gradually giving way to intensive practices such as market gardening and arboriculture. "These three production systems have emerged in the east of the Cape Verde peninsula, thanks to the proximity of the Senegalese capital, to the detriment of rain-fed crops and extensive livestock farming"[71] .

In the locality, this phenomenon is manifested in the cultivation of groundnuts and market gardening, both family and large-scale, which plays an essential economic role these days. "Market gardening is the most innovative activity in urban and peri-urban agriculture. It provides a high income from very small areas, and is well suited to the food demands of urban dwellers. However, it generally comes up against a health problem: the produce is not safe to eat because of the use of contaminated water for irrigation"[72] .

4.1.2. In the livestock sector

In terms of livestock production, the poultry sector is booming, reflecting changing dietary habits and the presence of a market supported by neighboring towns such as Dakar, Thiès and Mbour. Since the advent of avian flu in 2005, which led to the closure of borders to imports of chicken thighs, the local poultry industry has filled up, with total sales now estimated at one hundred and twenty-eight (128) billion CFA francs (around $256 million).

According to Mbodj, local chick production more than doubled from six (6) million to seventeen (17) million between 2005 and 2012, while egg production rose from three hundred and twenty-four (324) million to six hundred and seventy-two (672) million. Poultry meat production rose from nine (9) tonnes to twenty-six (26) tonnes, while feed production increased from one hundred and two thousand two hundred and ninety (102,290) tonnes to one hundred and ninety-seven thousand eight hundred and sixty-three (197863) tonnes over the same period[73] .

---

[71] Diongue (M), 2010, op. cit. p. 53.

[72] Dauvergne (S), 2011, Les espaces urbains et péri-urbains à usage agricole dans les villes d'Afrique sub-saharienne (Yaoundé et Accra) : une approche de l'intermédiarité en géographie, Thèse de doctorat en Géographie, Defended 08-12-2011 à Lyon, École normale supérieure, p. 61.

[73] Mbodj (A-M), (2017), "Rapport sur l'aviculture au Sénégal", published on IPSNA (Inter Press Service News Agency), accessed August 17, 2021, https: //www.hubrural.org//

It's an activity that requires little space, as is already in short supply in the area, and plays a remarkable economic role. It is mainly practiced in highly urbanized localities and those bordering communication routes such as the RN1 and the toll freeway (Bentégné, Sindia, Boukhou, at the airport resettlement site, etc.).

"There's nothing we can do about it. The airport has taken all our homes and our land, which is ideal for farming and raising livestock. Since the resettlement sites are so narrow, we're forced to raise poultry at home to meet some of our financial needs. A good part of our production is sold in Sébikhotane or Diamniadio [...]. The state has deprived us of our activities without supporting us". [74]

The other phenomenon worth noting is the development of the sheep sector, which accounts for 8.5% of the area's livestock production, with the integration of new, improved breeds, better known as luxury breeding. In the absence of space, a large proportion of the population is involved in this activity, especially in the more urban areas of Sindia, Ndayane, Popenguine, Guéréo and Diass, where the dominant breed is the *Ladoum*[75] . "For more than three decades, sheep farming, especially of imported breeds (*Ladoum, Bali-Bali, Azawatt*, etc.), has been gaining ground in Senegal's major cities. [...]. Some breeders keep sheep called *Ladoum*. The breeding of *Ladoum* sheep, considered by some as a prestige activity, is becoming increasingly attractive to certain socio-professional categories"[76] .

In contrast to the cattle sector, much of the feed is purchased or produced, which has led to an increase in groundnut cultivation in the area in recent years. The intensive nature of peanuts, and above all their financial value, means that they are more demanding in terms of feed, health and hygiene.

---

[74] Semi-structured interview with Ibrahima Pouye, resident of the airport resettlement site in the commune of Keur Moussa, conducted on September 10, 2022.

[75] The Ladoum sheep is a hypermetric, lanky animal. Its average height at the withers is 105 $\pm$ 3.56 cm for males and 88.8 $\pm$ 6.11 cm for females. Body length is 93.5 $\pm$ 2.08 cm for males and 83.2 $\pm$ 8.07 cm for females. The dominant coat color is black and white. Females often have strong horns and udders.

[76] Fall (A-K), Dieng (A) and Ndiaye **(S)**, 2017, L'élevage des moutons de race Ladoum dans la commune de Thiès, Sénégal : caractéristiques socioéconomiques et techniques, *Afrique SCIENCE* 13(4) (2017) 140 - 150, ISSN 1813-548X, http://www.afriquescience.info, p. 141.

These are species that are highly sensitive to climatic variations and irresistible to water and food deficits. This is why the species are put in very good conditions, with good financial cover, to ensure good production. "For sheep farms with between 3 and 10 sheep, 10 and 20 sheep and more than 20 sheep, feed costs represent 93.42%, 92.36% and 77.68% of expenses respectively. Annual gross margins for sheep farms vary from 270,000 to 1,444,400 FCFA"[77] .

Photo 3: *Ladoum* sheepfold

Photo 4: Image of a *Ladoum* male

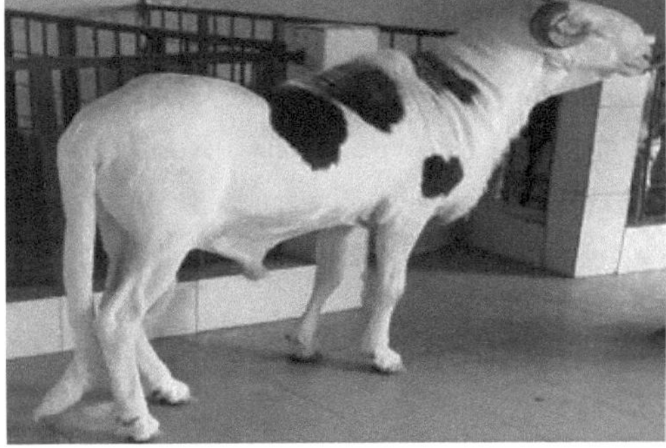

Aside from the network of breeders who sell their products to each other, much of the production is sold during family ceremonies or religious events such as the Tabaskie, and showrooms are located in urban areas or on major thoroughfares for greater visibility.

4.2. Other business sectors

---

[77] Fall (A-K), Dieng (A) and Ndiaye **(S)**, 2017, op.cit., p.147

The most important phenomenon observed among this population is their reconversion. Lacking the space to carry out their activities, many agropastoralists have converted to embrace other activities as a way of adapting to and keeping pace with the changes taking place in the area.

Table 7: Population share of business sectors

Secteurs	Types d'activités		Fréquence
Agropastoralisme	Agriculture et élevage		**52%**
Activités extra-agropastorales	La maçonnerie	**12,9%**	**48%**
	Le gardiennage	**7,4%**	
	Le carrelage	**4,0%**	
	La menuiserie	**3,4%**	
	Le courtage	**3,4%**	
	Les employés industriels	**6,1%**	
	Les employés agricoles	**10,2%**	
	Autres	**0,6%**	
Total			**100%**

Source: Field surveys, 2022

Analysis of these data shows a proliferation of non-agricultural occupations, with a boom in land and building-related occupations (masonry 12.9%), and agricultural employees given the development of peri-urban and commercial agriculture, with 12.2% of the population. Industry is also well represented, with 6.1% of the population, and the main units are Ciments du Sahel (CDS) and *Twyford Ceramics,* not forgetting the mining sector, which is concentrated in the Diass commune (Togglou, Bandia, Tchiky) and employs many people. "The commissioning of the quarries has above all enabled the development of migration flows of skilled workers and day laborers into the rural community"[78] .

Despite these alternatives, the retraining of the population remains problematic, as a large proportion of them find it difficult to fit in with these companies, given their low level of training or education.

4.3. Limitations of these alternatives

It should be noted that the majority of workers have little or no expertise or qualifications, often unsuitable for positions of responsibility within these

---

[78] Ndiaye (S), 2012, *État des Lieux de l'Environnementale et des Ressources Naturelles de la Communauté Rurale de Sindia,* Geography dissertation, UCAD, p.45.

structures. Thus, "The rapid urbanization of sub-Saharan Africa is uncoupled from a supply of stable, salaried jobs"[79] .

What's more, demand far outstrips supply, which is why we don't decide what we're going to do or how much we're going to get. The urban dynamics of Dakar, characterized by "sustained demographic growth and economic performance insufficient to satisfy the high demand"[80] , are at the root of this phenomenon.

All these factors, combined with the loss of land, put agropastoralists at a disadvantage, forcing them to take on whatever activities are available to support their families. These activities are often carried out under less than ideal conditions, with little remuneration.

The above-mentioned factors are at the root of the unprofitability of some of these activities in the area. But it should also be noted that: "The densification of the urban fabric is not the result of an industrial boom as in the West, but has led to the hypertrophy of this sector as a result of a process that has been carried out without major guidance, has often been poorly managed and has failed to provide basic social services"[81] .

4.3.1. Profitability of these activities

A large proportion of the population (38.7%) think that these activities are moderately profitable, against 31.2% who maintain the opposite. The proportion of the population who think that these activities are not very profitable is estimated at 13.6%, and at 16.3% for those who admit that they are very profitable. Assessing the profitability of these activities makes it possible to distinguish between the level of qualification and responsibility of the workers. Those who think that the profitability is "very good", constitute the educated class and the base of the company, with a salary level above 100,000 Fcfa. This is a very small group, and they are often employed on a permanent basis. And the other extreme, made up of the mass who think that

---

[79] Simonneau (C), 2015, *Gérer la ville au Benin, La mise en œuvre du Registre foncier urbain à Cotonou, Porto-Novo et Bohicon*, Thèse présentée à la Faculté de l'aménagement en vue de l'obtention du grade de Ph.D. en aménagement, Université de Montréal, academia.edu, p. 17.

[80] Sidibé (M), 2006, *Observatoires de développement local, Dialogue politique sur la production sociale de l'habitat*, Edition Adis, Enda Thiers monde, Dakar, p. 9.

[81] Gueye (F-N-D) et *al*, 2009, op. cit. p. 13.

these activities are "unprofitable", represents the workers in the true sense of the word, without qualifications. This group has only its own strength to offer, and is often indentured at the mercy of others, with derisory salaries that never reach 100,000 Fcfa, and is employed on a daily basis.

Figure 5: Profitability of these activities

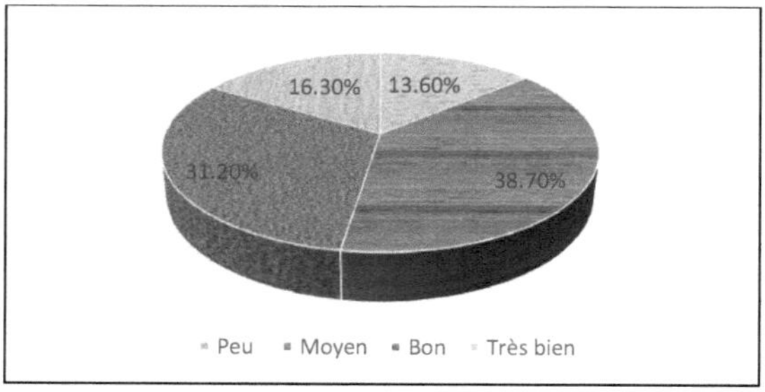

Source: Field surveys, 2022

So we need to understand that the level of qualification is an important barometer that can dictate stability within a company, but also consideration from a salary point of view. Unfortunately, a large proportion of the population who are thinking of retraining do so under conditions that are not conducive to employability or the enhancement of human dignity.

4.3.2. Salary levels

This population is mainly made up of seasonal and day laborers, more often women and retired people. "The income thus generated contributes to improving the living conditions of women, who are potential actors in development, by taking care of their families and children"[82] . The Dutch company *Van Oers,* based in Kiréne *and* active in market gardening, beats the record with almost 4,000 jobs, the majority of which are held by women. And according to Séne (2013), in his dissertation work, carried out on this farm, wages are low. "Nearly 73% of employees earn incomes of no more than 40,000 CFA francs. It also shows that 13.74% earn between 40 and 60,000

---

[82] Thiandoum (M), 2011, "Potentiel de production fruitière et économie villageoise en pays Saféne", *Annale de la FLSH*, n°41/B, p. 73-74

CFA francs. And those earning more than 100,000 CFA francs account for 1.92%"[83].

This situation is further proof that this population has relatively low monthly incomes. This often places them in a precarious situation, bordering on the deleterious, with serious problems in meeting family needs. Despite their diversity, these alternative activities fail to feed their men. Nearly 64.6% of this population believe that the income from these activities is insufficient to cover their needs, against 35.3% who maintain the opposite.

Figure 6: Monthly salary levels in Fcfa

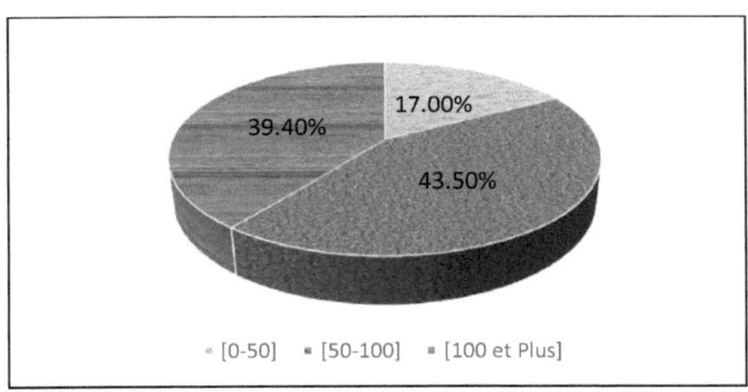

Source: Field surveys, 2022

"We can't help it, and the work is very hard (7 a.m. to 6 p.m.). What's more, we foreigners have to pay rent and food, and the rest is sent home to our parents. And we don't have a union, or a unitary framework that can defend our interests, and if you make a demand you're sent back without rights"[84].

This phenomenon can be explained on the one hand by the generally difficult economic context, characterized by high demand for jobs compared to very low supply. Another is the elasticity of families in rural areas, where all family members live under the same roof and share responsibilities.

4.3.3. Capacity to cover requirements

Faced with low wages and a high cost of living, heads and fathers are forced to combine their activities with others, to diversify their income. This

---

83 Séne (M), 2013, Analyse des impacts socio-économiques du Périmètre maraicher *Van Oers* dans la Rurale de Diass, Mémoire master 2 géographie, FLH, UCAD, 100 pages.

[84] Semi-structured interview with O. W, an employee of the Chinese factory (*Twyford Ceramics*), located in the commune of Sindia, conducted on September 10, 2022. He doesn't want to risk losing his job, so he asked us to abbreviate his name and we respected the survey protocols.

phenomenon is much more prevalent among the indigenous population, and especially at certain times of the year (during the rainy season or events).

It should be noted that these difficult conditions, especially in urban centers such as the capital, have prompted a good proportion of the population to leave these localities and make a definitive return to the area, given the rise of new trades. For most of them, the lack of profitability of their small businesses and their lack of expertise may also be factors encouraging their return. They are welcomed by these companies to earn a living, but often on a part-time basis.

Figure 7: Capacity to cover needs

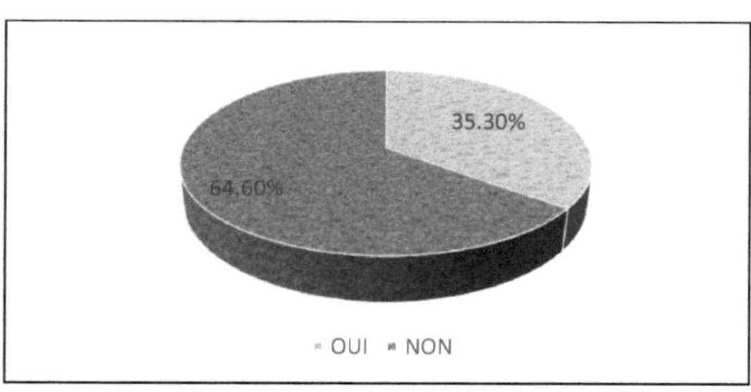

Source: Field surveys, 2022

Low wages justify their inability to cover the needs of the population. This is perceived by 64.60% of the population, versus 35.30% who think the opposite. As a result, a large proportion of the population (46.7% vs. 53.3%) combine their activities with others to increase their income and provide for their families. Complementary activities are mainly cash crop farming, in localities with some land reserves such as Thiafoura, Kiréne, Bandia, and the targeted varieties are eggplant, okra, cassava, groundnuts.In addition to the extraction sector, other crops include cassava, sorrel, etc., poultry farming and small-scale trading, especially along the RN1[85] , in the Sindia, Boukhou, Bentégné and Diass localities.

This situation is more common among workers who are not employed full-time by these companies, or when their own business is struggling. It should

---

[85] Route Nationale1, linking Dakar to the tourist town of Mbour, passing through the Saféne region (from Diass to Sindia).

be noted that these complementary activities are characterized by their ephemeral and uncertain nature, which often compromises their profitability.

Conclusion

With the reduction in agropastoral areas, caused by urbanization of the area and building development, these activities will be in a very unfavorable situation, close to decadence. This phenomenon will force those involved in the sector to retrain as a means of resilience.

Firstly, in the agropastoral sector, with the development of cash crop activities such as market gardening, peanut cultivation and agribusiness by major investors, and the poultry sector, which is more common in localities close to communication routes. The sheep sector is beginning to take off, especially in urban areas, in the form of luxury breeding, with improved species such as the *Ladoum*. This was followed by the development of extra-agropastoral activities, made up of the informal sector, characterized by new industrialization, trade, transport, etc., and captivating many workers.

But it should be noted that the profitability of these activities is not so desirable, as a large proportion of this population has a relatively low salary level, given their low level of expertise. As a result, their income will be unable to cover their family needs, prompting them to combine it with other activities to make up the shortfall.

GENERAL CONCLUSION

Given its geographical position, the Dakar-Thiès-Mbour triangle is undergoing major changes in the direction of urbanization. The advance of the urban front, driven mainly by the capital to relieve congestion, is constantly torpedoing the saféne country. It should be noted that the urbanization of this locality is first and foremost a matter of state will, not only with its attempt to decongest the capital, but also to correct special inequalities, while planning this locality, considered to be the antechamber of the capital and the future nerve center of Senegal's economy. This aspiration of the State will be materialized by the implementation of many structuring infrastructures, which have contributed enormously to the reshaping of the morphology of this locality.

This desire on the part of the State was later supported by the national and international private sector, with a great deal of investment in many areas, given the locality's favorable territorial offer. These investments are often punctuated by industrial infrastructures, which diversify the area's economy by employing a large number of people.

It should be noted that these dealings are not without consequences for the daily lives of the population, particularly in terms of land, with a staggering decline in areas suitable for crops and pasture. Many of the infrastructure projects in the area, such as the AIBD and its related maneuvers, are land-intensive. In addition to the land they occupy, these infrastructures continue to attract national and foreign investors in search of space for their projects.

As a result, many areas have left the hands of agropastoralists, to the benefit of the state and investors, dealing a devastating blow to these traditional activities, the backbone of the local economy. This has led to a disorientation of these activities in terms of surface area, assets and income, resulting in a significant drop in yields, to the benefit of a growing informal sector. Faced with this situation, agropastoralists are looking for a happy way out, trying to reconvert to other trades as a way of resilience.

The problem is that these activities are not very profitable, and pay relatively low wages. It should be noted that they are carried out in deleterious conditions, closer to human alienation than to moral satisfaction. Low wages can be explained by a superiority of demand over supply, but also by the population's lack of preparation in terms of qualifications. The other phenomenon that may explain this situation is the elasticity of families in rural areas, where a large proportion of them live under the same roof and each is called upon to contribute according to his or her physical or financial means. With this in mind, we need to ask ourselves whether we should sacrifice this locality's agro-pastoral potential in favor of urbanization, given that for several decades now, Senegal, through its public policies, has been on a hellish quest for food sovereignty, knowing that this locality, from BUD Sénégal to GOANA, via REVA, has always been the receptacle for these large-scale projects?

BIBLIOGRAPHICAL REFERENCES

ANAT, (2015), Schéma directeur d'aménagement et de développement territorial de la zone Dakar-Thiés-Mbour, Ministère de la Gouvernance Locale, du Développement et de l'Aménagement du Territoire, Rapport provisoire, 163 pages.

ANSD, (2012), Situation économique et sociale du Sénégal en 2012, Ministère de l'économie des finances et du plan, Division de la Documentation, de la Diffusion et des Relations avec les Usagers ISSN 0850-1491, 15 p.

Ba Awa and Moustier Paul, (2010), Perception de l'agriculture de proximité par les résidents de Dakar, *Revue d'Économie Régionale & Urbaine,* N°5, Edition Armand Colin, pages 913-936.

World Bank, (2015), IBRD-IDA, *Urbanization Review:* Emerging cities for an emerging Senegal, 126 pages.

World Bank, (2009), World Development Report.

Bertrand Monique, (1994), *La question foncière dans les villes du Mali, Marchés et Patrimoines,* Publié avec le concours du CNRS, KARTHAL-ORSTOM, page. 11.

Bernus Emond and Boutrais Jean, (1994), Crises et enjeux du pastoralisme africain, Géographes de l'ORSTOM, département MAA, 213 rue La Fayette, 75480 Paris Cedex 10. CR. Acad. Agrlc. Fr, 80, n° 8, p.109.

Bocquier Philippe, (1999), "La transition urbaine est-elle achevée en Afrique?", *Chronique du CEPED*, n°34, p. 1-4.

Chaleard Jean Louis, and Dubresson Alin, (1999), *Villes et campagnes dans les pays du Sud : géographies des relations*, Paris, Karthala, p.59.

Da Cunha Antonio and Matthey Laurent, (2007), "La ville et l'urbain : des savoirs émergents*", Presses polytechniques et universitaires romande*, p. 25.

Dauvergne Sarah, (2011), Les espaces urbains et péri-urbains à usage agricole dans les villes d'Afrique sub-saharienne (Yaoundé et Accra) : une approche de l'intermédiarité en géographie, Thèse de doctorat en Géographie, Defended 08-12-2011 à Lyon, École normale supérieure, p. 61.

Djah Armand Joséo *et al,* (2017), "Les implications socio-économiques et spatiales des pratiques foncières des autorités coutumières dans le développement de la ville de Yamoussoukro (Côte d'Ivoire)", *Regardsud*, N°2, page 163.

Demoulin Dominique, (1970) *Étude géomorphologique du massif de Diass et de ses bordures (Sénégal occidental,* PhD thesis, 3rd cycle in Geography, UCAD, p.37.

Diong Momar, (2010), *Périurbanisation différentielle : mutations et réorganisation de l'espace à l'est de la région dakaroise (Diamniadio, Sangalkam et Yéne), Sénégal*, Thèse de doctorat (type de thèse) de Géographie, encadreur (dir.), Université Paris Ouest Nanterre La Défence, p. 14-16.

Diop Amadou, (2011), *Aménagement en Zone littorale et risques industriels : l'espace Thiaroye sur mer-Mbao,* PhD thesis, UCAD, FLSH, Department of Geography, 332 p.

Diop Mariame, (2016), *La contribution des programmes immobiliers du pôle urbain de Diamniadio dans la résolution de la crise du logement à Dakar*, ESES ex ENEA, Mémoire de master 2, ESEA ex ENEA, Aménagement du territoire, environnement et gestion urbaine (ATEGU), p.122.

Doriel Apprili Élisabeth, (2006), *Les Ires grandes villes dans l'urbanisation*, Institut français de l'urbanisme, p.76.

Faye, Bernard, (2006), "Les pastteurs sont des éleveurs " contemplatifs ". In Courage G. (dir), l'Afrique des idées reçues. Paris: Belin, coll. "Mappemonde", 399 pages.

Fall Abdou Khadre, Dieng Abdoulaye and Ndiaye Saliou, (2017), L'élevage des moutons de race Ladoum dans la commune de Thiès, Sénégal : caractéristiques socioéconomiques et techniques, *Afrique SCIENCE* 13(4) (2017) 140 - 150, ISSN 1813-548X, http://www.afriquescience.info, p. 141.

Gueye Ndéye Fatou Diop et al, (2009), *Agriculteurs dans les villes Ouest africaines ; Enjeux fonciers et accès à l'eau,* IAGU-KARTHALA-CREPOS, p.191.

Houndechandji Marie François, (2012), *Urbanisation et gestion des ressources naturelles dans le domaine des Niayes à Dakar : impacts sociaux, environnementaux et sanitaires*, Thèse de doctorat unique, UCAD, département de géographie, p. 10.

Jacob -Jean-Pierre and Delville Philippe Lavigne, (1994), *Les associations paysannes en Afrique : organisation et dynamiques*, APAD-KARTHALA-IUED, p. 293.

Ka Ibrahima, (2018), Enjeux et défis de la gouvernance foncière en Afrique : contribution à l'étude du droit foncier à partir d'une analyse de sécurisation foncière rurale au Burkina Faso et au Sénégal, PhD Thesis in Law, Université Gaston Berger de Saint Louis, p.137.

Landais e. and Lhotse ph, (1990), "L'association agriculture élevage en Afrique intertropicale", *in* Sociétés pastorales et développement" Paris, *Cah. Sc. Hum*, ORSTOM, 26, 1-2, 217-235.

Marty André, (1993), "La gestion des terroirs et les éleveurs: un outil d'exclusion ou de négociation", *in* Revue Tiers-Monde, Paris, PUF, T.XXXIV, n° 134, 327-344.

Mayke Kaag, Yaram Gaye and Marieke Kruis, (2013), "Les conflits fonciers au Sénégal revisités: continuités et dynamiques émergentes", in *Gerti Hesseling, A l'ombre du droit,* p.38.

Mbodj Amadou Moukhtar, (2017), "Rapport sur l'aviculture au Sénégal", published on IPSNA (Inter Press Service News Agency), accessed August 17, 2021, https: //www.hubrural.org//

Morgan Davide L, (1996), Focus Group, Annual Review of Sociology, 22, 129-152, http: //dx.doi.org/10.1146/annurev.soc.22.1.129.

United Nations, (2004), World Urbanization Prospects.

Ngana François et *al*, (2010), *Transformations foncières dans les espaces périurbains en Afrique centrale soudanienne*, p.10.

Ndiaye Souleymane, (2012), *État des Lieux de l'Environnementale et des Ressources Naturelles de la Communauté Rurale de Sindia,* Geography dissertation, UCAD, p.45.

Olivier de Sardan Jean Pierre (1995) "La politique du terrain", Enquête [En ligne], 1 | 1995, online July 10, 2013, accessed October 15, 2022, France. URL : http://enquete.revues.org/263 PEDIDAS http://www.pdidas.org/fr.

Piermay Jean Luck et *al*, (2007), *La ville sénégalaise. Une invention aux frontières du monde*, Paris, Karthala, 242 p.

Pouye Issa, (2003), *La communauté rurale de Diass : Etude géographique*, DEA de Géographie, encadreur, Faculté des Lettres Sciences Humaines, UCAD/Dakar, p.12.

Puépi Bernard, (2015), Les gouvernances foncières et leur impact sur le processus de développement : Cas de quelques pays africain, Paris, L'Harmattan, p.189.

UNDP, (2014), *Strengthening the resilience of urban agricultural systems: assessing urban and peri-urban agriculture in Dakar, Senegal*, https://attachment.outlook.live.net, 59 pages.

Séne Moussa, (2013), Analyse des impacts socio-économiques du Périmètre maraicher Van Oers dans la Communauté Rurale de Diass, Mémoire master 2 géographie, FLH, UCAD, 100 pages.

Seck Souleymane, (2011), *Analyse des mutations foncières au Sénégal: CAS de la communauté rurale de Diass,* Master 2 thesis, UCAD, geography department, GDER, p. 59.

Sidibé Mariéme, (2006), *Observatoires de développement locale, dialogue politique sur la production sociale de l'habitat,* Edition Adis, Enda Thiers monde, Dakar, p. 9.

Simonneau Claire, (2015), *Gérer la ville au Benin, La mise en œuvre du Registre foncier urbain à Cotonou, Porto-Novo et Bohicon*, Thèse présentée à la Faculté de l'aménagement en vue de l'obtention du grade de Ph.D. en aménagement, Université de Montréal, <u>academia.edu</u>, p. 17.

Sow Mamadou, (2010), *Agglomération dakaroise au tournant du siècle, Vers une réinvention de la ville africaine?* PhD thesis in spatial planning - urbanism, Université Paris Ouest Nanterre La Défense, Ecole doctorale, économie, organisation et société, <u>academia.edu</u>, p. 25.

Tall Abdoulaye, (1999), "Croissance urbaine et stratégies résidentielles des ménages : l'exemple des quartiers spontanés de Dakar", *L'Union pour l'Etude de la Population Africaine*, n° 39, p. 7.

Thiandoum Mariama, (2013), *Espace périurbain et activité rurale: dynamiques territoriales en cours en pays saféne*, Thèse de Doctorat unique de Géographie, UCAD/ETHOS, 324 p.

Thiom Seybatou, (2017), Analyses des mutations foncières dans la commune de Sindia, Master 2 thesis, Geography, FLSH, UCAD, 118 pages.

TABLE OF CONTENTS

yes
I want morebooks!

Buy your books fast and straightforward online - at one of world's fastest growing online book stores! Environmentally sound due to Print-on-Demand technologies.

Buy your books online at
www.morebooks.shop

Kaufen Sie Ihre Bücher schnell und unkompliziert online – auf einer der am schnellsten wachsenden Buchhandelsplattformen weltweit! Dank Print-On-Demand umwelt- und ressourcenschonend produzi ert.

Bücher schneller online kaufen
www.morebooks.shop

Printed by Books on Demand GmbH, Norderstedt / Germany